DATE DUE

Aircraft Carriers
Supplies for a City at Sea

Multiplying Multidigit Numbers with Regrouping

John Strazzabosco

PowerMath™

The Rosen Publishing Group's
PowerKids Press™
New York

Published in 2004 by The Rosen Publishing Group, Inc.
29 East 21st Street, New York, NY 10010

Book Design: Ron A. Churley

Photo Credits: Cover, pp. 4–5, 24–25, 25 (inset), 26–27 © George Hall/Corbis; pp. 6–7 © Corbis; pp. 8–9,
11 (inset), 14–15, 30 © AFP/Corbis; pp. 10–11, 22–23 © Reuters NewMedia Inc./Corbis; pp. 13, 20 by
Leonello Calvetti, Lorenzo Cecchi, Luca Massini, Donato Spedaliere; pp. 14 (left inset), 16 (inset), 20 (inset)
© Yogi, Inc./Corbis; p. 14 (right inset) © Hartwell Thomas/Corbis Sygma; pp. 16–17 © U.S. Navy/Corbis;
pp. 18–19 © Corbis Sygma; pp. 28–29 © Jacques Langevin/Corbis Sygma.

Library of Congress Cataloging-in-Publication Data

Strazzabosco, John.
   Aircraft carriers, supplies for a city at sea : multiplying multidigit
numbers with regrouping / John Strazzabosco.
        p. cm. — (PowerMath)
Includes index.
Summary: Explains how to multiply multidigit numbers by exploring modern
aircraft carriers and their crews.
   ISBN 0-8239-8995-X (lib. bdg.)
   ISBN 0-8239-8919-4 (pbk.)
   6-pack ISBN: 0-8239-7447-2
   1. Multiplication—Juvenile literature. 2. Aircraft
carriers—Juvenile literature. [1. Multiplication. 2. Aircraft
carriers.] I. Title. II. Series.
   QA115 .S83 2004
   513.2'13—dc21
                                    2002155343

Manufactured in the United States of America

Contents

The Navy and Its Aircraft Carriers   5

The USS *Enterprise*   8

A City at Sea   12

Comfort on an Aircraft Carrier   19

Getting the Job Done   22

A Dangerous Job   27

Glossary   31

Index   32

The Navy and Its Aircraft Carriers

Think of the United States military as a tree. The army, navy, and air force are branches. The army has soldiers on land. The air force operates planes in the air. The navy sails ships on the water. However, the army does have some boats, and the navy has some planes.

The navy has aircraft carriers for its planes. An aircraft carrier is a huge navy ship. It carries up to 85 planes and **helicopters**, and can hold over 6,000 people! About half of these people run the ship, and the other half run the planes. Planes can land on the carrier deck and take off from it.

The aircraft carrier deck is like a navy base with an airport. The navy can sail this base and airport to any sea or ocean. This makes the aircraft carrier very useful.

The United States spends about $5 billion when it builds a new aircraft carrier! That is a lot of money, but it means that the United States and other countries are well protected.

If 12 U.S. aircraft carriers each bring 85 planes to the coast of Greenland, how many U.S. planes can fly over Greenland from the aircraft carriers?

$$
\begin{array}{r}
\overset{1}{85} \text{ planes per carrier} \\
\times\ 12 \text{ carriers} \\
\hline
\overset{1}{170} \\
+\ 85 \\
\hline
1,020 \text{ planes}
\end{array}
$$

Sometimes other countries may not want U.S. planes flying over their land or U.S. soldiers on their land. However, the navy can sail its airport anywhere there is an ocean, because oceans are considered **international waters**. Anyone can use oceans because no country owns them. An aircraft carrier can usually bring a navy base and its airport close to where they are needed.

Floating an airport around the world takes many people. The aircraft carrier might need to stay at sea for months or even years at a time. Members of the ship's navy crew live on board. They need lots of radios, phones, and computers. They also need doctors, dentists, and things like toothbrushes, soap, and books. They can usually find what they need right on the aircraft carrier.

The United States has about 12 aircraft carriers, making it the world's largest fleet. Some countries have only 1 aircraft carrier.

The USS *Enterprise*

Aircraft carriers get their names in different ways. The USS *George Washington* is named after the first president of the United States. The USS *Kitty Hawk* was named after the town where the Wright brothers made the first successful plane flight in 1903. Every aircraft carrier's name starts with the initials "USS." This stands for "United States Ship."

The very first United States aircraft carrier was called the USS *Langley* in honor of a man named Samuel Langley, an American scientist who performed early flight experiments. The carrier went into service in 1922. The aircraft carrier USS *Enterprise* was built in 1961. It was the world's first **nuclear**-powered aircraft carrier. The word "enterprise" means a willingness to take part in a daring action. When working at **full capacity**, the *Enterprise* can carry about 5,800 people and 85 planes.

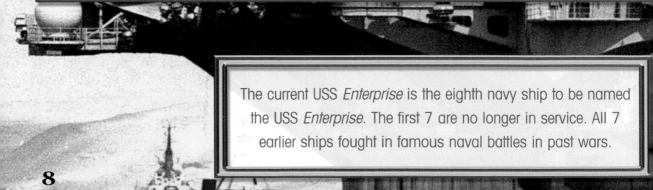

The current USS *Enterprise* is the eighth navy ship to be named the USS *Enterprise*. The first 7 are no longer in service. All 7 earlier ships fought in famous naval battles in past wars.

9

The navy plans for each aircraft carrier to last 50 years. Since the USS *Enterprise* was built in 1961, that means it should be retired—or go out of service—by 2011. For now, though, the USS *Enterprise* is still in operation.

Older aircraft carriers are called light carriers. Newer ones are called supercarriers. The supercarriers are larger than light carriers and are able to stay at sea longer. The USS *Enterprise* is a supercarrier.

Supercarriers began to be built in the 1950s, when jet planes replaced **propeller** planes. Jet planes are heavier, use more fuel, and land at a faster speed than the propeller planes did. The design of the supercarrier reflected the navy's changing needs.

Every aircraft carrier has a tall tower called an "island." This is the command center for all flight operations and for the entire ship.

If 12 U.S. aircraft carriers each have 589 women on board, how many women are on all 12 carriers?

$$
\begin{array}{r}
\overset{1\ 1}{589} \text{ women per carrier} \\
\times\ \ \ 12 \text{ carriers} \\
\hline
\overset{1\ 1}{1\ 178} \\
+\ 5\ 89 \\
\hline
7{,}068 \text{ women}
\end{array}
$$

command center

A City at Sea

Aircraft carriers are like floating cities. Each aircraft carrier has an airport right on its deck. From top to bottom, an aircraft carrier can stand as high as a 25-story building. Each supercarrier is about 1,100 feet long. That's longer than 3 football fields! With a crew of about 6,000 people and with 85 planes, a carrier is larger than many towns on land. A completed carrier can weigh more than 95,000 tons!

These large "floating cities" need many supplies to operate. People on board the carrier must have food to eat and water to drink. There are also other everyday problems to consider. A person might get a toothache and need a dentist, or become ill and need to see a doctor. The navy has thought about such problems. Each aircraft carrier has all the supplies and services a person needs while at sea.

The supercarrier USS *John F. Kennedy* is as long as the Empire State Building is tall! It was named after the thirty-fifth president of the United States.

USS *John F. Kennedy*

Empire State Building

Toothaches do happen to sailors. Let's say 55 teeth are pulled on an aircraft carrier each month. If so, how many teeth get pulled on the aircraft carrier each year?

¹
55 teeth pulled each month
x 12 months in a year
110
+ 55
660 teeth pulled in a year

health club

post office

Like the USS *Enterprise,* the USS *Kitty Hawk* can also carry about 85 planes. Up to 5,600 people can live on board. All these people need many services and supplies to stay healthy and live comfortably.

On board the carrier USS *Kitty Hawk* you might find 4 doctors and 5 dentists, as well as 65 hospital beds and 6 operating rooms. If you needed to buy anything, you could go to one of the USS *Kitty Hawk's* 4 stores. The carrier has a laundry, a fire department, and 2 barbershops that give an average of 27,000 haircuts each year! The USS *Kitty Hawk* also has a bakery, a bank, a library, and a museum, as well as its own radio and television stations and its own daily newspaper.

The USS *Kitty Hawk* even has its own health club and its own post office.

How many soft drinks do the people aboard the USS *George Washington* drink in an average 2-week period?

$$\begin{array}{r} \overset{3}{1{,}900} \\ \underline{\times\ 14} \\ \overset{1}{7\ 600} \\ +\ 19\ 00 \\ \hline 26{,}600 \end{array}$$

1,900   **soft drinks per day**

x 14   **days**

7 600

+ 19 00

26,600   **soft drinks**

kitchen

Like the USS *Kitty Hawk*, the USS *George Washington* is also able to take care of all the people who live on board. In an average year, the carrier's doctors perform about 100 **surgeries** and give over 250 physical exams to the aircraft carrier's pilots. The carrier's **pharmacists** fill about 16,800 orders for medicine. Nurses give over 11,000 shots!

An aircraft carrier's crew gets hungry and thirsty, too. The nearly 6,000 sailors on a carrier eat the following in an average day:

18,000 meals   900 fresh loaves of bread
500 gallons of milk   1,900 soft drinks
11,000 eggs

All the food served aboard an aircraft carrier has to be bought, stored, and cooked. Just 2 hamburgers for each sailor for dinner would require 12,000 burgers in 1 day!

mess hall

If 452 sailors each drink 16 ounces of fresh water a day, how many total ounces do they drink?

```
      3 1
      452 sailors
   x   16 ounces a day
   1
   2 712
 + 4 52
   7,232 ounces a day
```

Comfort on an Aircraft Carrier

Because so many people live on an aircraft carrier, making sure they all have fresh water to drink is important. An aircraft carrier is surrounded by the ocean's saltwater, but people cannot drink saltwater.

Like other aircraft carriers, the USS *Enterprise* makes its own fresh water from saltwater. Each year, millions of gallons of saltwater are made into fresh water for people to drink. This process happens in a **desalinization** plant. First, the saltwater is boiled. The steam boils off and is collected. The salt is left behind. The steam, or vapor, is then cooled down and becomes fresh water.

Using desalinization plants aboard aircraft carriers means that they don't have to find a way to store large amounts of water for long periods of time. When more fresh water is needed, it can be made right on board!

On an aircraft carrier, the kitchen is called the "galley" and the dining room is called the "mess hall." Each carrier has several galleys and several mess halls.

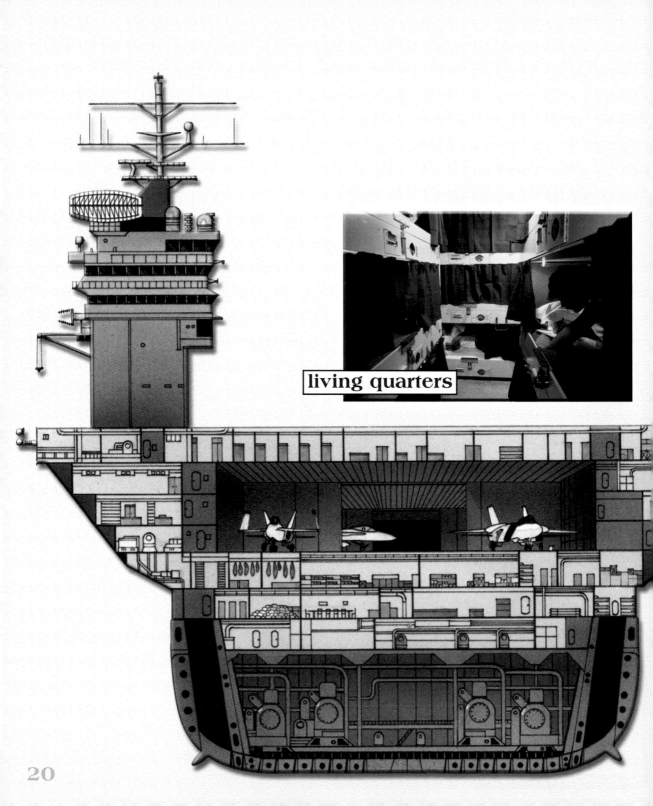

living quarters

The process of desalinization also makes a lot of heat, which can present a major problem. The rooms aboard an aircraft carrier must be kept cool enough to be comfortable for people to live in. About 60 sailors share a single sleeping compartment and bathroom, and bunkrooms next to the desalinization plant can get very hot. The engineers and maintenance crew must solve these kinds of problems.

On the supercarrier USS *Harry S. Truman*, there are air conditioning plants to keep the ship cool and the sailors comfortable. These air conditioning plants have to keep the carrier's rooms cool so the sailors can sleep at night. In fact, an air conditioning system on an aircraft carrier is powerful enough to cool almost 3,000 homes!

This diagram shows a cross section of the USS *Harry S. Truman*, which was commissioned in 1998. The tower to the left is the "island," or flight command center.

Getting the Job Done

Because aircraft carriers are like floating cities, they have their own zip codes. People aboard the carrier can send mail. They can also get mail from their friends and family. The mail plane lands right on the aircraft carrier to bring the mail. Up to 1,500 pounds of mail are processed each day!

It is a difficult job to land a plane on deck because the aircraft carrier is hard to see from the air. It is painted gray so that it blends in with the ocean when seen from the air. That is done on purpose so enemy pilots can't easily find the carrier. Sailors are always busy touching up the gray paint. The carrier stores hundreds of gallons of paint and lots of paintbrushes and rollers.

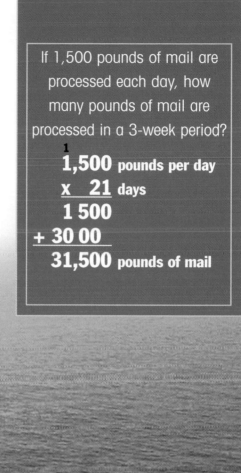

If 1,500 pounds of mail are processed each day, how many pounds of mail are processed in a 3-week period?

$$\begin{array}{r} 1 \\ 1{,}500 \text{ pounds per day} \\ \underline{\times \quad 21} \text{ days} \\ 1\ 500 \\ \underline{+\ 30\ 00} \\ 31{,}500 \text{ pounds of mail} \end{array}$$

The color of an aircraft carrier is called "haze-and-gray." Hiding a ship by using color is called "camouflaging."

On the deck of an aircraft carrier, events happen fast. Pilots need to know quickly who is doing what, and so does the captain of the carrier. Every job has a team of people to do it. The different teams wear different colored vests. Here are the colors of vests and what they mean:

Yellow:	These people position the planes on deck.
White:	These people are safety inspectors.
Brown:	These people are **maintenance** workers.
Blue:	These people drive the tractors that pull the planes.
Green:	These people operate the **catapult** and cables.
Purple:	These people fuel the planes.
Red:	These people load **missiles**, bombs, and rockets.
Silver:	These people are part of the fire and rescue team.

Jets are sent out on missions almost every day. During times of war, flight operations run for about 18 hours each day!

The fuel supply tank holds 50,000 gallons. There are 68 planes to refuel. Let's say each plane needs 669 gallons of fuel for a mission. Is there enough fuel for the mission?

$$
\begin{array}{r}
\overset{4\ 5}{\underset{5\ 7}{}} \\
669 \text{ gallons} \\
\times\ 68 \text{ planes} \\
\hline
5\ 352 \\
+\ 40\ 14 \\
\hline
45{,}492 \text{ gallons}
\end{array}
$$

The 50,000-gallon supply tank will be able to do the job.

plane refuelers

If the USS *George Washington* travels all day, every day, how far will it travel in 1 week at 34 miles per hour? First, figure out how many hours are in a week. Then, multiply that number by the number of miles traveled each hour.

$$\begin{array}{r} \overset{2}{24} \text{ hours} \\ \times\ 7 \text{ days} \\ \hline 168 \text{ hours in a week} \end{array}$$

$$\begin{array}{r} \overset{2}{\underset{}{\overset{2}{2}}}\ \overset{2}{\underset{}{\overset{3}{2}}} \\ 168 \text{ hours} \\ \times\ 34 \text{ miles per hour} \\ \hline \overset{1}{672} \\ +\ 5\ 04 \\ \hline 5{,}712 \text{ miles} \end{array}$$

A Dangerous Job

The carrier USS *George Washington* can travel at 30 knots, or about 34 miles per hour. The planes go much faster. A catapult is used to get the planes into the air. The planes zip along the deck from zero to 165 miles per hour in just 2 seconds!

Taking off is risky, but so is landing on the carrier deck. Cables are stretched across the deck to catch the landing planes. When a plane is landing, it lowers a hook, and the hook catches onto one of the cables. This stops the plane quickly. If the plane misses that cable, it tries to catch one of the other cables. If that fails, a safety net at the end of the deck catches the plane.

Once the plane's hook catches onto the cable on deck, the plane travels about 300 feet before it comes to a stop.

28

The deck on some carriers is turned at a slight angle. The runway points slightly toward the side of the ship. This is done so that when a plane doesn't stop in time, it will hit the safety net instead of crashing into the control tower or into other planes on deck.

There are other dangers that must be considered aboard an aircraft carrier. Jet fuel can explode. The carrier also stores bombs, rockets, and missiles. With so many risky supplies on board, people must be very careful to keep the carrier safe. This requires a lot of planning, and the navy must do its homework. They start planning a new carrier 10 years before it is even built!

Parked planes are not left on the deck for long. They are carried below deck by an elevator. They are then stored below deck until their next takeoff.

The United States's newest supercarrier, the USS *Ronald Reagan,* went into service in 2003. This aircraft carrier is named after former U.S. president Ronald Reagan, who served from 1981 to 1989.

Like the navy's other aircraft carriers, the navy plans for this new supercarrier to last for 50 years. It will also be supplied with many items and services for its large crew, making it another of the navy's cities at sea.

Glossary

catapult (KA-tuh-puhlt) A machine that pushes a plane along the runway to help it take off.

desalinization (dee-sa-luh-nuh-ZAY-shun) The process of removing the salt from saltwater to make fresh water.

full capacity (FUL kuh-PA-suh-tee) Using the total amount of space.

helicopter (HEH-luh-kahp-tuhr) An aircraft that has large, spinning blades on top to raise the aircraft and make it fly.

international waters (in-tuhr-NA-shuh-nuhl WAH-tuhrz) Large bodies of water, like oceans, that belong to no nation.

maintenance (MAYN-tuhn-uhns) The process of keeping something in good working condition.

missile (MIH-suhl) A kind of rocket that delivers a bomb to a target.

nuclear (NOO-klee-uhr) Using atomic energy. Atomic energy is energy that exists inside an atom, one of the tiny bits of matter that make up all things.

pharmacist (FAR-muh-sist) Someone who is trained to give medicines to people when their doctor suggests it.

propeller (pro-PEH-luhr) A metal blade that turns rapidly to make an aircraft move.

surgery (SUHR-juh-ree) A medical operation.

Index

B
bakery, 15
bank, 15
barbershops, 15
bombs, 24, 29

D
dentist(s), 7, 12, 15
desalinization, 19, 21
doctor(s), 7, 12, 15, 17

F
fire and rescue team, 24
fire department, 15
fuel, 10, 24, 29

L
Langley, Samuel, 8
library, 15
light carriers, 10

M
maintenance workers, 24

N
newspaper, 15
nuclear-powered, 8
nurses, 17

P
pharmacists, 17

R
radio and television stations, 15
Reagan, Ronald, 30

S
safety inspectors, 24
stores, 15
supercarrier(s), 10, 12, 30

U
USS *Enterprise*, 8, 10, 15, 19
USS *George Washington*, 8, 17, 27
USS *Harry S. Truman*, 21
USS *Kitty Hawk*, 8, 15, 17
USS *Langley*, 8
USS *Ronald Reagan*, 30

W
water, 12, 19
Wright brothers, 8